Science Close-Up
CRYSTALS

Written by Robert A. Bell
Illustrated by Paul Lopez

A GOLDEN BOOK • NEW YORK
Western Publishing Company, Inc., Racine, Wisconsin 53404

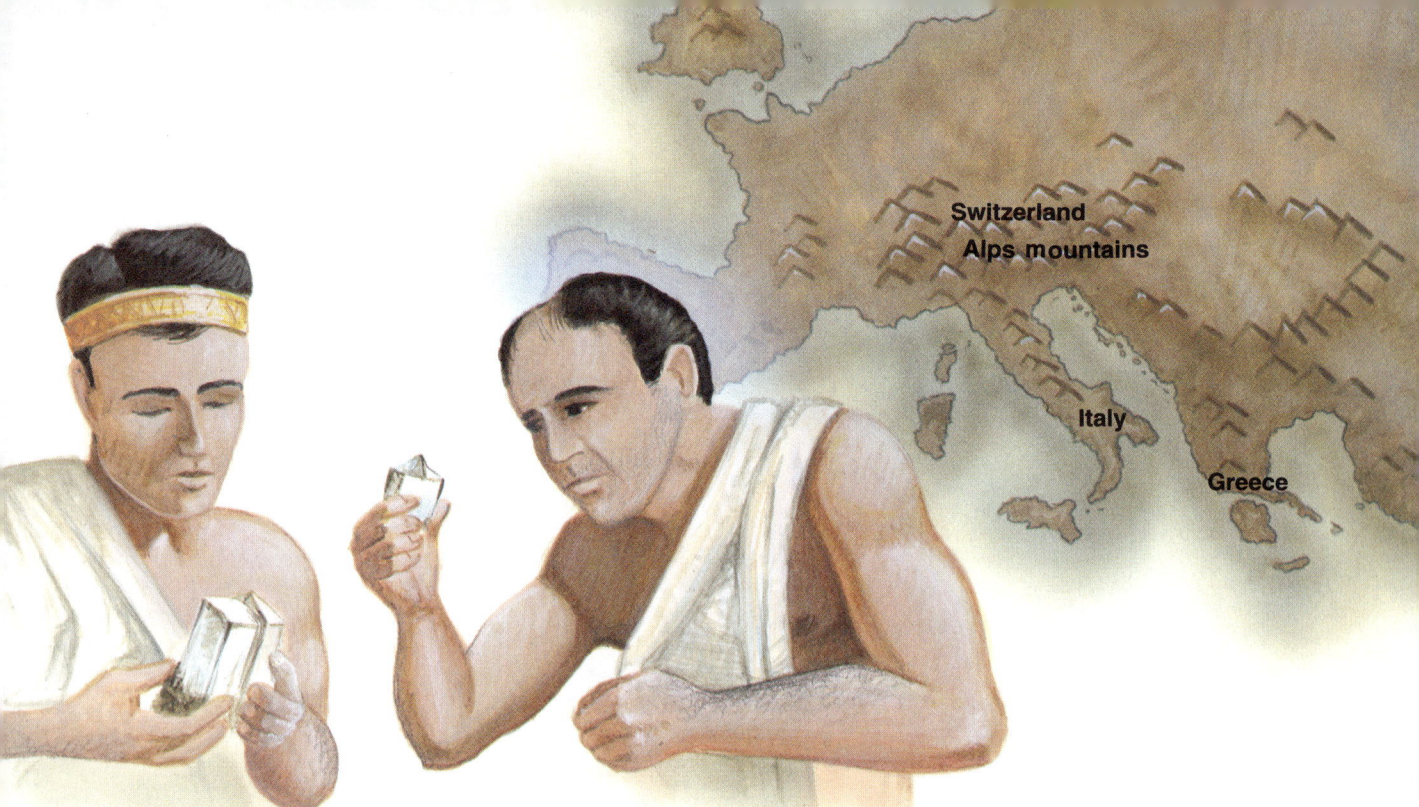

What Is a Crystal?

A long, long time ago, people from Greece crossed the Alps mountains in central Europe. There they found something beautiful and strange.

It was a group of solid, heavy stones. Unlike other stones, these were clear and colorless and sparkled brightly in the sunlight. The Greeks called the stones *krystallos*, meaning "ice." They thought the stones were pieces of ice— frozen so hard, they would never melt. Today we know the stones were not ice at all. They were *crystals* of **quartz**, just one of thousands of crystals known today.

What is a crystal? A crystal is a rock with flat faces, or surfaces. Many crystals pulled from the ground look as if they have been cut and polished. But they haven't been. The crystals *grew* that way.

Atoms are the tiny pieces from which everything in the world is made. In a crystal, molecules—which are groups of atoms—join together and line up in ways that make the crystal faces flat.

clear quartz crystals

salt crystals

Making Crystals

You can see how crystals grow in your own home.

In a glass or cup, add 1 teaspoon of salt to 3 tablespoons of water. Stir until all of the salt is dissolved. Then pour the salted water onto a flat-bottomed pan or cookie sheet. Leave the pan overnight or longer. When the water has evaporated, you will find a white crust. Look at the crust through a magnifying glass and you will see it is made up of tiny square shapes. These are crystals of salt.

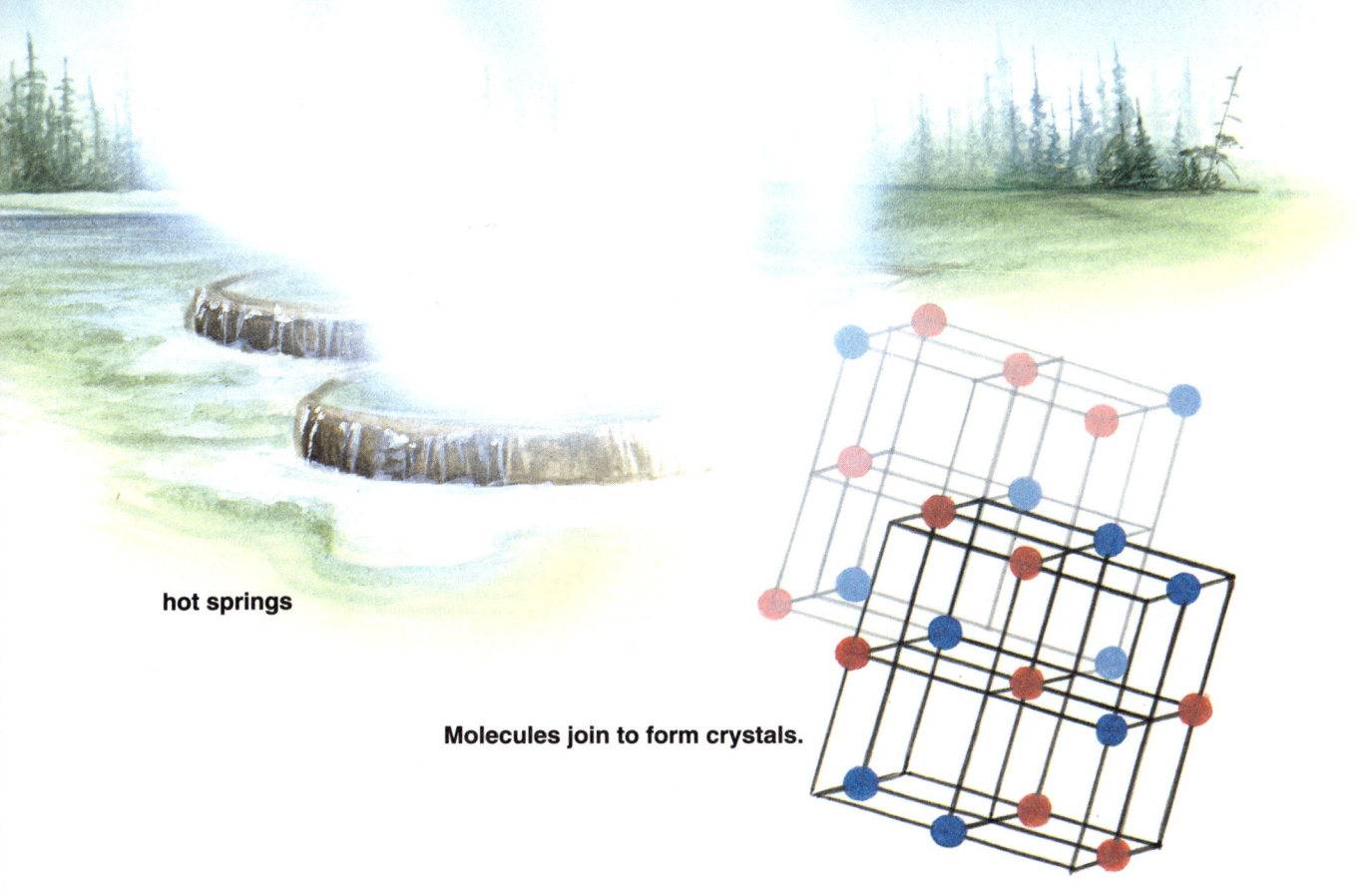

hot springs

Molecules join to form crystals.

In nature crystals form in the same way. A crystal of quartz starts out as many atoms and molecules mixed up in hot water or melted rock underground. Eventually the melted rock gushes out of the ground as volcanic lava and starts to cool. The heated water rushes out as a hot spring, then turns into steam and evaporates in the cooler air.

As they cool, the atoms and molecules of quartz attract each other like magnets—just as the molecules of salt attracted each other in your pan. The quartz molecules join together in rows, lining up over and over again in the same order. After a long time, the molecules make crystals.

crystal chandelier

workman pouring glass

glass goblet

What a Crystal Isn't

Clear, sparkling, perfect glass is often called "crystal." But expensive, high-quality glassware and glass chandeliers are only glass, and glass is not a crystal. Glass does not grow into shapes with flat faces. Glass does not grow at all. It is made by people. To make glass, sand is melted in a very hot fire until it becomes a liquid. Then the liquid is poured into a mold. When the liquid cools, it turns into hard, clear glass.

A strange thing about glass is that even after it hardens it can act like a liquid. Glass can slowly bend and flow like water. Look at old windows and you can see that the glass is no longer flat. It is rippled. Over time, the glass has flowed downward and become thicker at the bottom than at the top.

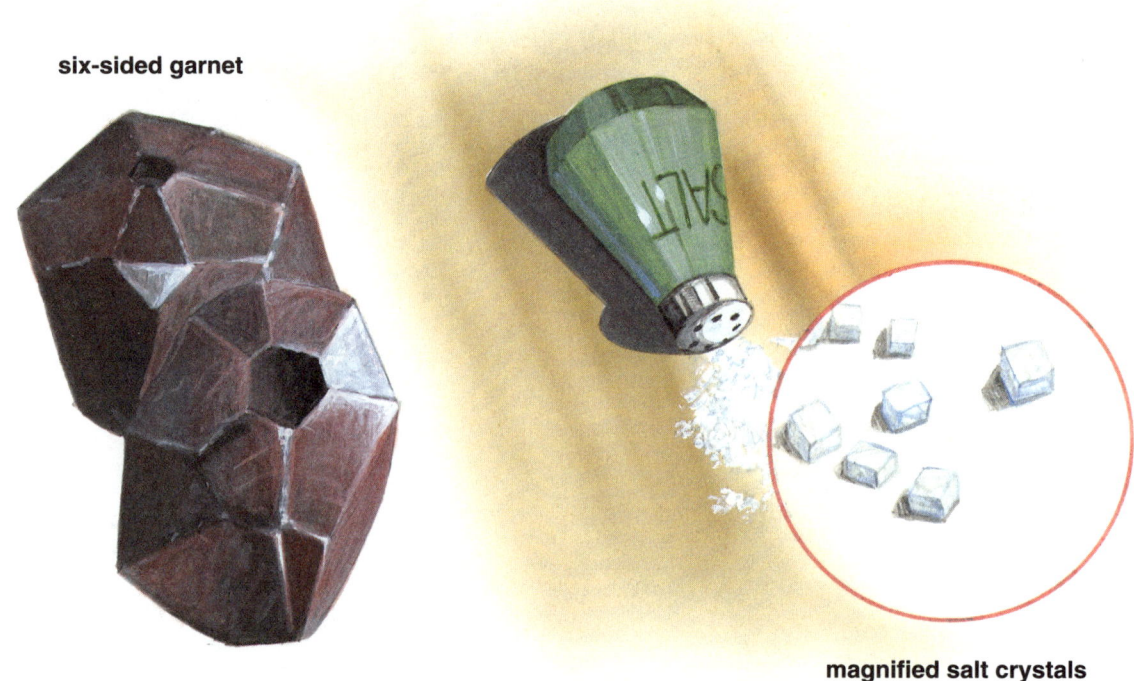

six-sided garnet

magnified salt crystals

Crystal Shapes

When a crystal of quartz grows, it almost always makes a six-sided shape—like a pencil. But other kinds of crystals grow into other shapes. Salt is made up of tiny cubes. That's the shape of salt crystals—and can be the shape of **garnets**, too. These shapes can be seen by the naked eye or examined more closely with a magnifying glass. In fact, each kind of crystal has a shape that it takes most of the time. A crystal gets its shape from one of the six ways its molecules can line up.

Crystal shapes are not always easy to see. Sometimes crystals press against other rocks as they grow, which changes their shape. Sometimes two crystals grow into each other to form what are called "twins." Crystals that tumble along the bottom of a river may be rubbed round and smooth, losing their original shape. Then it takes an expert to tell exactly what kinds of crystals they are.

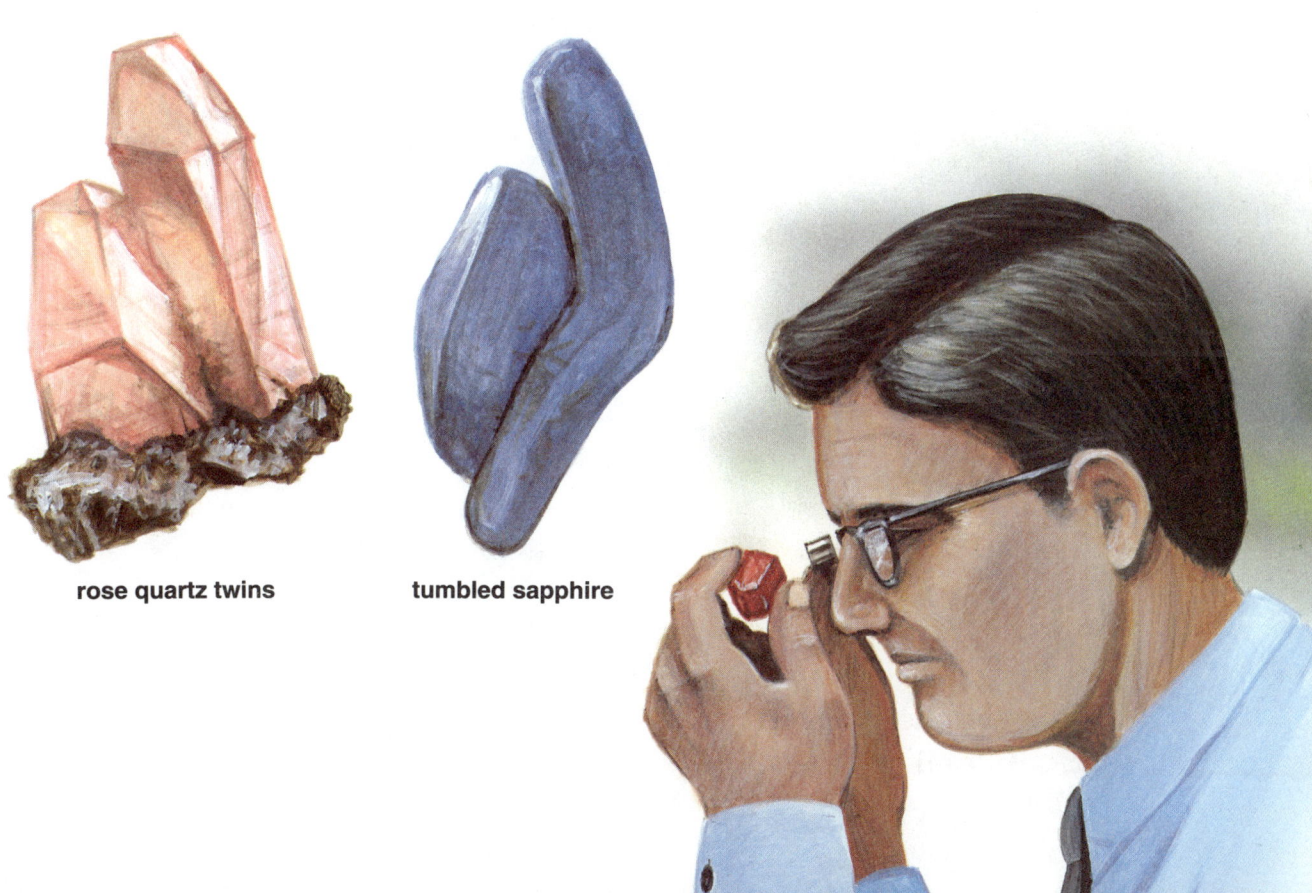

rose quartz twins   tumbled sapphire

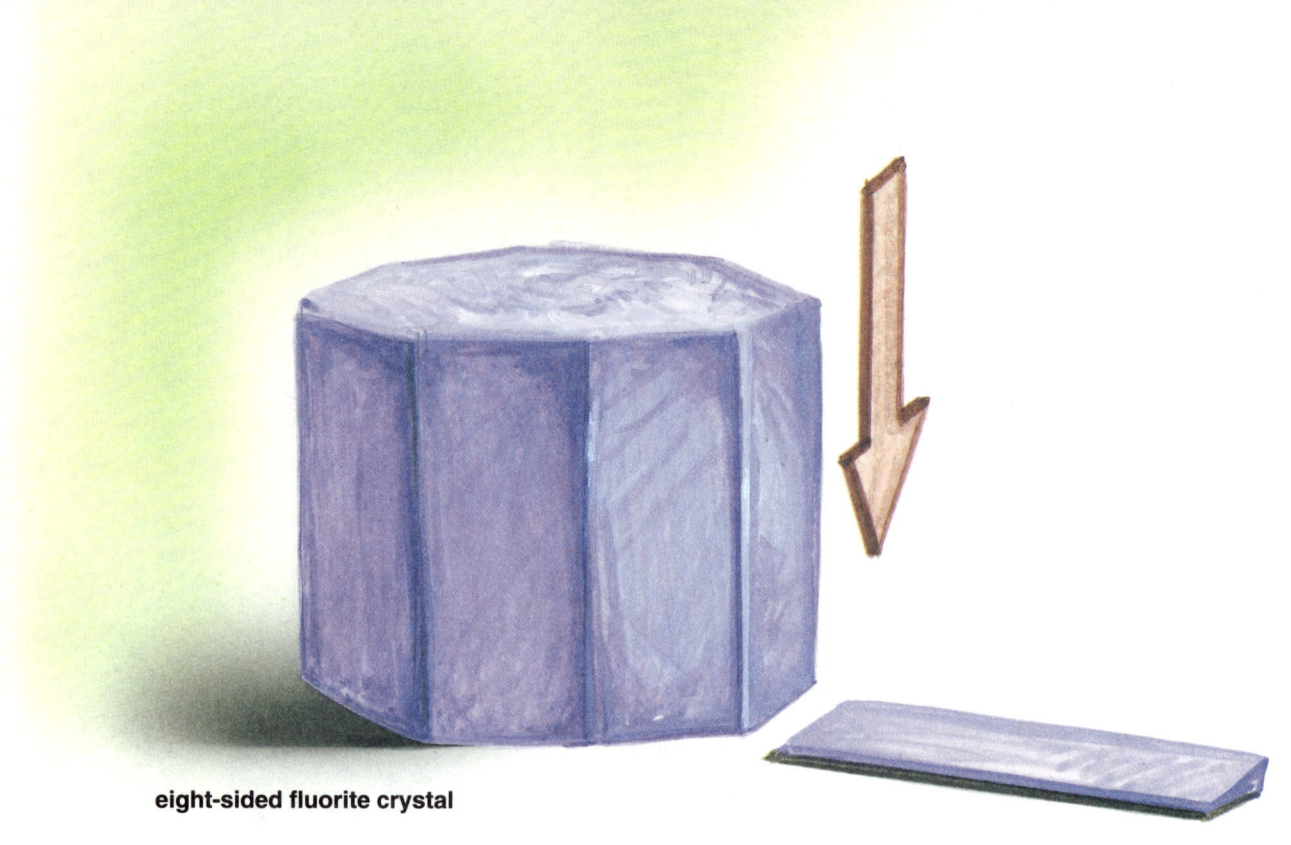

eight-sided fluorite crystal

Crystal Cleavage

If you hit a crystal of **fluorite** in exactly the right place, it breaks easily. And the break will have a smooth, flat surface. The ability of crystals to break into smaller pieces with flat sides is called "cleavage." Even **diamonds**—which are very, very hard—will cleave, or break, into smooth pieces if you hit them in the right place.

Cleavage happens because of the regular lineup of atoms and molecules inside the crystal. This lineup also gives crystals their shape. Fluorite grows into shapes with eight sides, and it will cleave into smaller eight-sided shapes.

cleaving fluorite

Games With Light

Crystals shine. Crystals sparkle. The beauty of crystals comes partly from the "games" they play with light.

When light falls on a crystal, some light bounces from the outside of it back to your eye. Then you can see the outside of the crystal shine. At the same time, some light goes inside the crystal, where it bends and bounces before leaving the crystal. What you see then is a sparkle from inside the crystal. Inside many colorless crystals, white light breaks into the colors of the rainbow. The more a crystal bends and bounces light inside, the more beautiful it is.

Some crystals play special tricks with light. For example, if you look through a clear piece of a crystal called **calcite**, you see *two* of everything. Calcite splits light in half and sends two pictures to your eye.

calcite

peridot

red tourmaline

blue tourmaline

brown tourmaline

Crystal Colors

Some crystals come in more than one color—and that can make it hard to tell one crystal from another. Some kinds of crystal, like green **peridot,** are easy. They're always the same color. But **tourmaline,** for example, may be green, yellow, brown, red, blue, pink, black, purple—or even more than one color at a time. Watermelon tourmaline got its name because it is green on the outside and pink on the inside!

purple tourmaline

pink tourmaline

watermelon tourmaline

Like most crystals, tourmaline gets its color from metal dust that mixed with the crystal as it grew. The brown and blue colors come from iron. Manganese turns the crystal pink or red. Green comes from either chromium or vanadium.

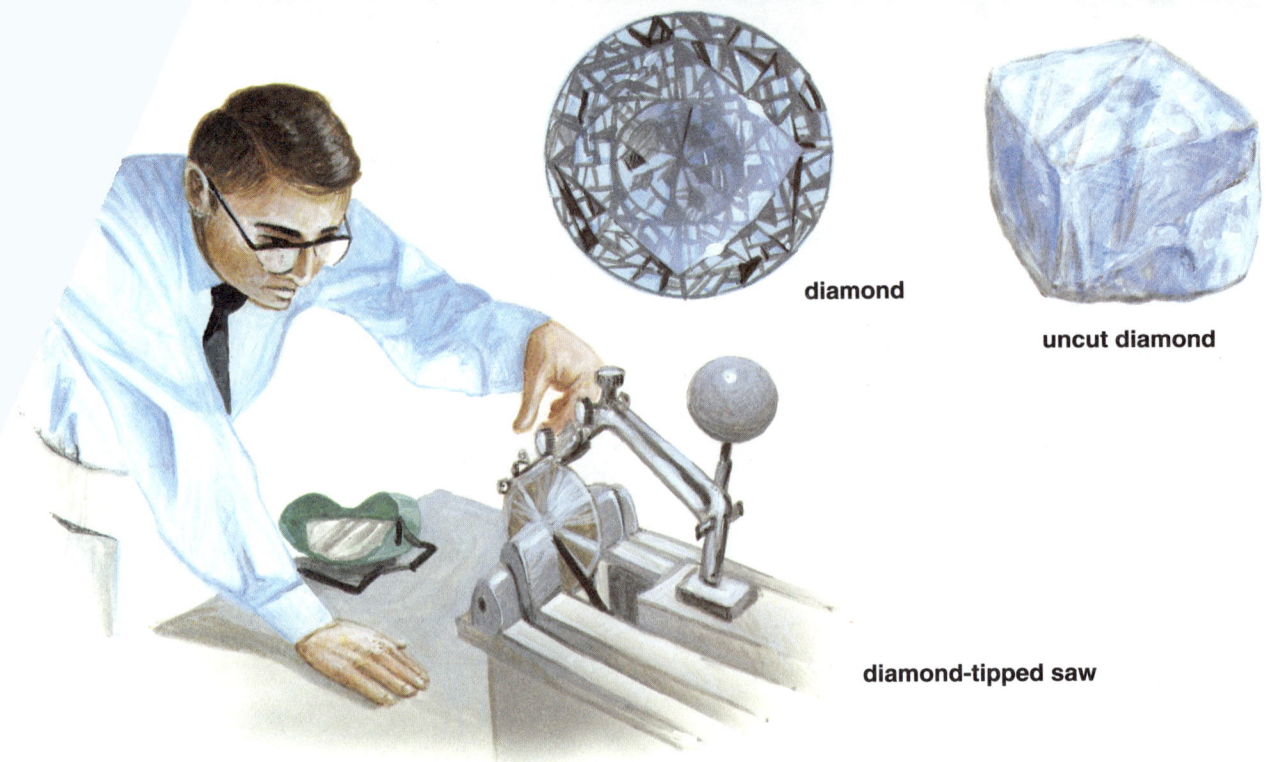

diamond

uncut diamond

diamond-tipped saw

Rare and Beautiful

The clearest and most sparkling crystals are called gems.
The most expensive of these are called precious gems.
Precious gems are valuable not only for their beauty, but
also because they are rare and hard to find.

Diamonds are the gems that everyone knows best.
They are the hardest natural substance on Earth.
Diamonds are so hard that they can only be cut by a saw
made of other diamonds! The most expensive diamonds,
the ones used in jewelry, are clear and colorless.
However, diamonds can also be yellow, blue, brown, pink,
and black.

Many people think diamonds are the most valuable gems, but **rubies** are more expensive—because they are harder to find. Rubies are red crystals of a mineral called corundum. Blue crystals of corundum are called **sapphires** and are among the most beautiful of gems.

Emeralds are bright green crystals of beryl. Beryl is a mineral that comes in many other colors as well. Light blue beryl is called **aquamarine**.

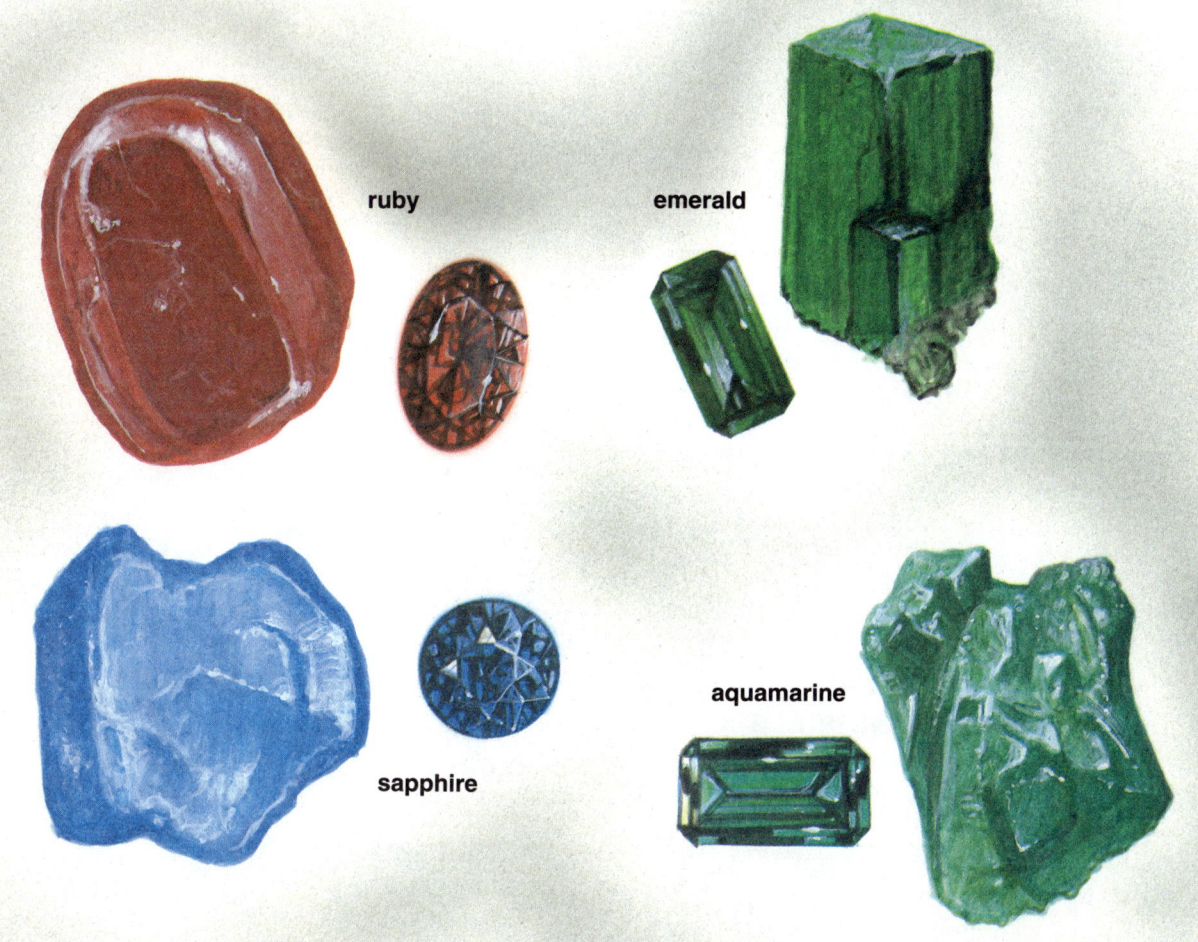

ruby

emerald

sapphire

aquamarine

clear quartz

rose quartz

amethyst

uncut clear quartz

citrine

Quartz and Garnet:
Useful and Colorful

Most gems are not as rare or costly as diamonds and
rubies, and are called semiprecious. There is more quartz
on Earth than any other mineral, and quartz gives us
many semiprecious gems. Quartz forms large, clear
crystals. When it is purple, quartz is called amethyst.
Pink crystals are called rose quartz, and yellow crystals
are called citrine.

Quartz is also a very useful crystal. All radios and TVs have thin pieces of quartz in them, which help tune in the picture or the sound.

Garnet is the name for a whole family of crystals used as semiprecious gems. Garnets are usually a strong, deep red in color, but they can also be reddish-purple, reddish-brown, orange, and even green.

uncut red garnet

red garnet

green garnet

reddish-violet garnet

orange garnet

red rose garnet

garnet

emerald

Crystal Magic

For thousands of years, people wanted to own crystals not only for their beauty, but also for the magic powers the stones were believed to hold.

Some people thought witches and wizards could see the future by looking into crystal balls made from quartz. Other crystals were supposed to protect people from sickness. Emeralds were called "healing stones." People wearing garnets were thought to be safe from wounds and poison. Diamonds were supposed to have power against poison, too, and ground-up diamonds were sometimes fed to sick people. Many people thought wearing sapphires would protect them from eye diseases.

What people believed about crystals often depended on where they lived. In Asia, people thought that wearing opals kept them healthy. But people in Europe thought opals brought bad luck. England's Queen Victoria, however, loved opals. She added them to the Royal Crown Jewels, and gave opal jewelry to her daughters and granddaughters as well.

opal necklace

Brazilian emerald mine

Finding Crystals

People find crystals by mining, or digging, into the earth.
Some mines are long, underground tunnels; others are
open pits. The miners collect dirt from the mines and sift
it carefully for valuable crystals.

People also hunt for crystals in rivers. As rivers flow
over the centuries, they slowly cut a deeper and deeper
riverbed. The river then picks up bits of dirt and rock
and tumbles them along the bottom. In places where
crystals are known to have been found, people search
river bottoms for loose gems.

Most of the world's crystals come from Brazil, South Africa, India, Sri Lanka, Burma, Thailand, and Australia. But the United States also has crystal treasures. Rubies and garnets are found in North Carolina, sapphires in Montana. Emeralds are mined in California, tourmaline in Maine. Crystals of quartz are found in nearly a dozen states.

South African open-pit diamond mine

hunting for rubies in Sri Lankan river

Crystals Under Glass

Many museums have fine displays of crystals large and small. Visitors can see the interesting shapes and fantastic colors of a number of different kinds of crystals. Some of the museums you can visit are listed below.

Arizona
Phoenix: Arizona Mineral Museum

Arkansas
Hot Springs: Quartz Crystal Cave and Museum

California
San Diego: San Diego Natural History Museum
San Francisco: Mineral Museum

Colorado
Denver: Denver Museum of Natural History

Georgia
Macon: Museum of Arts and Sciences

Illinois
Chicago: Field Museum of Natural History

Iowa
Waterloo: Museum of History and Science

Massachusetts
Cambridge: Harvard University Geological Museum

Minnesota
St. Paul: Science Museum

Missouri
Joplin: Tri-State Mineral Museum

New York
New York City: American Museum of Natural History

Ohio
Cincinnati: Cincinnati Museum of Natural History

Oregon
Portland: Oregon Museum of Science and Industry

Washington, D.C.
Smithsonian Institution